# Home Defense Book!

Laurel D. Malvern

Copyright and Legal Disclaimer:

© 2024 by Laurel D. Malvern. All rights reserved. No part of this publication may be reproduced, distributed, or transmitted in any form or by any means, including photocopying, recording, or other electronic or mechanical methods, without the prior written permission of the publisher, except in the case of brief quotations embodied in critical reviews and certain other noncommercial uses permitted by copyright law.

The information provided in this book, "Home Defense Book," is for general informational purposes only. While the author, Laurel D. Malvern, has made every effort to ensure the accuracy and completeness of the information presented, the author and publisher make no representations or warranties of any kind, express or implied, about the completeness, accuracy, reliability, suitability, or availability of the information contained within these pages.

Readers are advised to consult with professionals and experts in the relevant fields before implementing any of the strategies, techniques, or practices discussed in this book. The author and publisher disclaim any liability for any loss, injury, or damage incurred as a direct or indirect result of the use of the information provided in this book.

Furthermore, the author and publisher do not endorse any specific products, brands, services, or organizations mentioned within the text unless explicitly stated. Any reference to third-party products, services, or websites is provided for informational purposes only and does not constitute an endorsement or recommendation.

Readers are responsible for their own actions and decisions. The techniques, methods, and advice presented in this book are not guaranteed to be suitable or effective in all situations. Individual circumstances may vary, and readers should exercise caution and use their own discretion when applying the information contained herein.

By reading this book, readers acknowledge and agree to hold harmless the author, Laurel D. Malvern, and the publisher from any and all claims, liabilities, losses, damages, costs, or expenses arising from their use of or reliance on the information provided in this book.

For permissions requests, inquiries, or additional information, please contact the publisher.

"Home Defense Book"

Chapter: Introduction 8

Chapter Overview: What to Expect from This Book 9

Chapter: Sustainable Living 11

Chapter: Exploring Eco-Friendly Habits 13

Chapter: Practical Tips for Sustainable Living at Home 15

Chapter: Success Stories of Sustainable Living 17

Chapter: Electrical Systems and Power Tools 19

Chapter: Basics of Wiring, Circuitry, and Electrical Troubleshooting 21

Chapter: Guide to Selecting and Using Power Tools Safely and Effectively 23

Chapter: DIY Projects for Home Improvement and Maintenance 26

Chapter: Power Systems and Energy Independence 29

Chapter: Setting Up and Maintaining Home-Based Renewable Energy Systems 32

Chapter: Strategies for Achieving Energy Independence 35

Chapter: Successful Off-Grid Living and Sustainable Energy Projects 38

Chapter: Survival and Emergency Preparedness 41

Chapter: Creating a Comprehensive Emergency Preparedness Plan for Your Home and Family 44

Chapter: Essential Survival Skills 47

Chapter: Tips for Stockpiling Emergency Supplies and Creating a Survival Kit 50

Chapter: Handling Natural Disasters 53

Chapter: Preparing for Earthquakes, Hurricanes, Wildfires, Floods, and Other Natural Calamities 56

Chapter: Evacuation Planning, Sheltering in Place, and Community Response Strategies 59

Chapter: Post-Disaster Recovery and Rebuilding Efforts 62

Chapter: Water Supply and Management 65

Chapter: Methods for Ensuring a Reliable and Safe Water Supply at Home 68

Chapter: Rainwater Harvesting, Water Purification Techniques, and Water Conservation Strategies 71

Chapter: Managing Wastewater and Sewage in an Environmentally Responsible Manner 74

Chapter: Land Use and Sustainable Gardening 77

Chapter: Principles of Sustainable Gardening, Permaculture, and Organic Farming 80

Chapter: Creating a Resilient Garden that Can Withstand Various Environmental Challenges 83

Chapter: Integrating Livestock, Composting, and Natural Pest Control Methods into Your Gardening Practices 86

Conclusion: Recap of Key Concepts Covered in the Home Defense Book 89

Chapter: Encouragement to Take Action and Implement Sustainable Living Practices, Emergency Preparedness Measures, and Survival Skills 92

Chapter: Inspiring Stories of Home Transformation into Resilient, Self-Sufficient Havens 95

Chapter: Closing Thoughts from the Author 98

# Chapter: Introduction

Welcome to the Home Defense Book, your ultimate guide to sustainable living, emergency preparedness, and survival skills.

In a world filled with uncertainties, being prepared is not just a luxury—it's a necessity. Whether you're facing the threat of natural disasters, societal upheaval, or unforeseen emergencies, having the knowledge and skills to defend your home and ensure the safety of your loved ones is paramount.

This book is your comprehensive resource for all things related to home defense. From sustainable living practices that reduce your environmental impact to emergency preparedness strategies that will keep you one step ahead of any crisis, we cover it all.

Throughout these pages, you'll find practical advice, actionable tips, and expert insights designed to empower you to take control of your safety and security. Whether you're a seasoned prepper or someone just beginning to explore the world of self-reliance, there's something here for everyone.

So, as you embark on this journey through the Home Defense Book, I invite you to open your mind, absorb the information presented, and most importantly, take action. Together, let's build a future where every home is a fortress of security, sustainability, and peace of mind.

Let's get started.

# Chapter Overview: What to Expect from This Book

In this chapter, we'll provide you with a brief overview of the wide array of topics covered in the Home Defense Book. From sustainable living practices to effective land use strategies, we've compiled a comprehensive guide to help you defend your home and ensure the safety and well-being of your loved ones.

Primarily, we will delve into sustainable living practices. You'll learn how to reduce your environmental impact, conserve resources, and embrace eco-friendly habits that promote a more sustainable way of life.

Next, we'll explore electrical systems and power tools. Whether you're a seasoned DIY enthusiast or a novice homeowner, you'll gain valuable insights into wiring, circuitry, and the safe and effective use of power tools for home maintenance and improvement projects.

Survival and emergency preparedness are critical skills in today's uncertain world. In this book, you'll discover essential strategies for preparing for emergencies, including first aid techniques, shelter building, and stockpiling supplies to ensure you're ready for whatever may come your way.

Natural disasters can strike at any time, and being prepared can mean the difference between life and death. We'll provide you with practical guidance on how to handle hurricanes, earthquakes, wildfires, floods, and other natural calamities, as well as tips for evacuation planning and community response strategies.

Managing your water supply is another key aspect of home defense. We'll show you how to ensure access to clean, reliable water, including rainwater harvesting, water purification techniques, and water conservation strategies.

Finally, effective land use is essential for creating a sustainable and self-sufficient homestead. You'll learn how to maximize the productivity of your land for food production, gardening, and livestock, while also preserving natural resources and promoting biodiversity.

By the end of this book, you'll be equipped with the knowledge and skills necessary to defend your home, protect your loved ones, and thrive in any situation. So, without further ado, let's dive into the Home Defense Book and start building a safer, more sustainable future together.

# Chapter: Sustainable Living

In today's rapidly changing world, the importance of sustainable living practices cannot be overstated. As we face environmental challenges such as climate change, resource depletion, and pollution, it has become increasingly clear that we must take action to reduce our impact on the planet and ensure a better future for generations to come.

Understanding the importance of sustainable living practices in modern times is the first step toward creating positive change. Sustainable living encompasses a wide range of actions and behaviors aimed at minimizing our ecological footprint and promoting harmony between humans and the environment.

At its core, sustainable living is about living in a way that meets the needs of the present without compromising the ability of future generations to meet their own needs. It involves adopting practices that are environmentally friendly, socially responsible, and economically viable.

One of the key aspects of sustainable living is reducing waste. This includes recycling materials, composting organic waste, and minimizing the use of disposable products. By reducing waste, we can conserve resources, reduce pollution, and mitigate the impact of landfill waste on the environment.

Another important aspect of sustainable living is conserving energy. This can be achieved through energy-efficient practices such as using LED light bulbs, insulating homes, and choosing energy-efficient appliances. Additionally, embracing renewable energy sources such as solar, wind, and hydro power can help reduce reliance on fossil fuels and decrease greenhouse gas emissions.

Water conservation is also essential for sustainable living. By reducing water usage, implementing water-saving technologies, and harvesting rainwater, we can help ensure access to clean water for future generations and protect precious water resources.

Sustainable living also encompasses sustainable transportation practices, such as walking, biking, carpooling, and using public transportation whenever possible. By reducing reliance on fossil fuel-powered vehicles, we can decrease air pollution, reduce greenhouse gas emissions, and promote healthier communities.

In addition to environmental considerations, sustainable living also encompasses social and economic aspects. This includes supporting local businesses, promoting fair trade practices, and advocating for social justice and equity.

Overall, sustainable living is about making conscious choices that promote a healthier planet and a better quality of life for all. By understanding the importance of sustainable living practices in modern times and taking action to incorporate them into our daily lives, we can work together to create a more sustainable and resilient future.

# Chapter: Exploring Eco-Friendly Habits

In this chapter, we will delve into the various eco-friendly habits that can help individuals and families reduce waste, conserve energy, and minimize their environmental impact. By adopting these habits, you can play a significant role in promoting sustainability and protecting the planet for future generations.

Reducing Waste:

Embrace the "3 R's" of waste management: Reduce, Reuse, Recycle. Reduce waste by avoiding single-use plastics and disposable products whenever possible.
Practice mindful consumption by only purchasing items that you truly need and opting for products with minimal packaging.
Embrace the concept of "upcycling" by repurposing old items into new and useful creations.
Compost organic waste such as food scraps and yard trimmings to create nutrient-rich soil for gardening.
Conserving Energy:

Turn off lights, appliances, and electronics when not in use to reduce energy consumption.
Invest in energy-efficient appliances and electronics that carry the Energy Star label.
Seal gaps and cracks in windows and doors to prevent heat loss during colder months and reduce the need for heating.
Use programmable thermostats to regulate heating and cooling systems efficiently.

Minimizing Environmental Impact:

Choose sustainable transportation options such as walking, biking, carpooling, or using public transportation.
Support local farmers and businesses to reduce the carbon footprint associated with transportation and shipping.
Opt for eco-friendly cleaning products and personal care items that are free from harmful chemicals and toxins.
Reduce water usage by fixing leaks, installing water-saving devices, and incorporating water-efficient landscaping practices.
Embracing Sustainable Food Practices:

Reduce food waste by planning meals, buying only what you need, and properly storing perishable items.
Support sustainable agriculture practices by purchasing organic, locally sourced, and seasonal foods whenever possible.
Grow your own fruits, vegetables, and herbs at home using sustainable gardening techniques such as composting and companion planting.

By incorporating these eco-friendly habits into your daily life, you can make a positive impact on the environment and contribute to a more sustainable future for all. Remember, every small change counts, and together, we can create a healthier planet for generations to come.

# Chapter: Practical Tips for Sustainable Living at Home

In this chapter, we will provide you with practical tips for implementing sustainable practices at home, focusing on recycling, composting, and using energy-efficient appliances. These simple yet effective strategies can help reduce your environmental footprint and promote a more sustainable way of life.

Recycling:

Set up a designated recycling station in your home to collect recyclable materials such as paper, cardboard, plastic bottles, glass jars, and aluminum cans.
Familiarize yourself with your local recycling guidelines and ensure that you are sorting materials correctly to maximize recycling efforts.
Rinse out containers before recycling to prevent contamination and ensure they can be processed effectively.
Look for opportunities to reduce waste by choosing products with minimal packaging or opting for reusable alternatives.
Composting:

Start a compost pile or bin in your backyard to recycle organic waste such as fruit and vegetable scraps, coffee grounds, eggshells, and yard trimmings.
Layer green materials (such as kitchen scraps) with brown materials (such as leaves or newspaper) to create a balanced compost pile.

Turn or aerate the compost regularly to promote decomposition and prevent unpleasant odors.
Use finished compost to enrich soil in your garden, flower beds, or indoor plants, reducing the need for chemical fertilizers and promoting healthy plant growth.
Energy-Efficient Appliances:

When purchasing new appliances, look for models that carry the Energy Star label, indicating they meet strict energy efficiency guidelines set by the Environmental Protection Agency.
Choose appliances with advanced features such as energy-saving modes, programmable settings, and temperature controls to optimize energy usage.
Consider upgrading old, inefficient appliances with newer, more energy-efficient models to reduce energy consumption and lower utility bills over time.
Maintain appliances regularly by cleaning filters, coils, and vents to ensure they operate efficiently and last longer.

By incorporating these practical tips into your daily routine, you can make a significant difference in reducing waste, conserving resources, and minimizing your environmental impact. Remember, sustainable living is about making small, mindful choices that add up to create a healthier planet for future generations.

# Chapter: Success Stories of Sustainable Living

In this chapter, we will explore inspiring success stories of individuals and communities who have embraced sustainable living practices and made a positive impact on the environment and their quality of life. These real-life examples serve as proof that sustainable living is not only achievable but also rewarding and fulfilling.

The Johnson Family: Bea Johnson and her family are pioneers of the zero-waste lifestyle. By adopting the principles of reduce, reuse, recycle, and rot (compost), they have managed to produce only a small jar of trash per year. Their journey has inspired countless individuals worldwide to rethink their consumption habits and strive for a waste-free lifestyle.

The Green School: Located in Bali, Indonesia, the Green School is an innovative educational institution that promotes sustainability and environmental stewardship. The school's curriculum is centered around hands-on learning experiences that teach students about permaculture, renewable energy, and eco-friendly building techniques. Students are encouraged to develop solutions to real-world environmental challenges, empowering them to become future leaders in sustainability.

The Transition Town Movement: Originating in Totnes, England, the Transition Town movement is a grassroots initiative that aims to build resilient communities in the face of climate change and resource depletion. Transition Towns focus on local solutions to global challenges, such as promoting local food production, transitioning to renewable energy sources, and fostering a sense of community resilience and self-sufficiency.

The City of Freiburg, Germany: Known as Germany's "greenest" city, Freiburg has become a model for sustainable urban development. The city boasts extensive public transportation networks, pedestrian-friendly streets, and a commitment to renewable energy sources such as solar and wind power. Freiburg's sustainable initiatives have not only reduced carbon emissions but also improved the overall quality of life for residents.

The Tiny House Movement: Across the globe, individuals are downsizing their living spaces and embracing the tiny house movement as a way to reduce their ecological footprint and live more simply. Tiny house dwellers prioritize minimalist living, energy efficiency, and self-sufficiency, often building their homes using recycled materials and incorporating off-grid systems such as solar power and rainwater harvesting.

These success stories demonstrate that sustainable living is not only feasible but also desirable and beneficial. By taking inspiration from these examples and incorporating sustainable practices into our own lives, we can create a more resilient, harmonious, and sustainable future for ourselves and generations to come.

# Chapter: Electrical Systems and Power Tools

In this chapter, we will provide an overview of electrical systems and safety precautions for homeowners. Understanding how electrical systems work and knowing how to use power tools safely are essential skills for maintaining a safe and functional home environment.

Understanding Electrical Systems:

Electrical systems in homes consist of wiring, outlets, switches, circuit breakers, and other components that deliver electricity to appliances, lights, and other electrical devices. The main electrical panel, also known as the breaker box or fuse box, distributes electricity throughout the home and contains circuit breakers or fuses that protect the electrical system from overloads and short circuits.
Electrical wiring is typically made of copper or aluminum conductors encased in insulation. Wiring is installed in walls, ceilings, and floors and connects to outlets, switches, and fixtures.
Safety Precautions for Homeowners:

Always turn off power at the main breaker or fuse box before performing any electrical work.
Use insulated tools and wear personal protective equipment (PPE), such as gloves and safety goggles, when working with electricity.
Never overload electrical circuits by plugging too many devices into a single outlet or using extension cords improperly.

Keep electrical cords and wires away from heat sources, water, and sharp objects to prevent damage and reduce the risk of electrical fires.

Regularly inspect electrical outlets, switches, and wiring for signs of damage or wear and tear. Replace any faulty components immediately.

Hire a licensed electrician for any electrical repairs or installations that are beyond your level of expertise.

Power Tools:

Power tools are essential for many home improvement and maintenance projects but can be dangerous if not used properly.

Before using a power tool, read the manufacturer's instructions and familiarize yourself with its features, functions, and safety precautions.

Always wear appropriate PPE, such as safety goggles, ear protection, and dust masks, when using power tools.

Inspect power tools before each use to ensure they are in good working condition. Check for loose or damaged parts, frayed cords, and missing safety guards.

Use power tools in a well-ventilated area and be mindful of potential hazards, such as flying debris, sharp blades, and rotating parts.

Keep power cords away from the cutting path and use cordless tools or extension cords equipped with ground fault circuit interrupters (GFCIs) when working in wet or damp conditions.

By understanding electrical systems and following safety precautions when using power tools, homeowners can help prevent accidents, injuries, and property damage. Always prioritize safety when working with electricity and power tools to maintain a safe and functional home environment.

# Chapter: Basics of Wiring, Circuitry, and Electrical Troubleshooting

In this chapter, we will cover the fundamentals of wiring, circuitry, and electrical troubleshooting. Understanding how electricity flows through a home's electrical system and being able to identify and resolve common electrical problems are essential skills for homeowners.

Wiring Basics:

Electrical wiring consists of conductors, typically made of copper or aluminum, that carry electricity from the main electrical panel to outlets, switches, and fixtures throughout the home.
Wiring is installed in walls, ceilings, and floors and is protected by insulation to prevent electrical shocks and fires. Different types of wiring are used for different applications, including branch circuits for outlets and lights, feeder circuits for large appliances, and service entrance conductors for incoming power from the utility.
Circuitry:

A circuit is a closed loop through which electricity flows from a power source (such as the electrical panel) to a load (such as a light or appliance) and back to the source.
Circuits are typically protected by circuit breakers or fuses that interrupt the flow of electricity in the event of an overload or short circuit.

Branch circuits serve specific areas of the home, such as bedrooms, kitchens, and bathrooms, and are protected by circuit breakers or fuses of appropriate size and rating.
Electrical Troubleshooting:

When troubleshooting electrical problems, it is essential to start by identifying the symptoms and determining the possible causes of the issue.
Common electrical problems include tripped circuit breakers, overloaded circuits, loose connections, and faulty outlets or switches.

Use a multimeter or voltage tester to check for the presence of voltage and continuity in electrical circuits and devices.
Inspect wiring, outlets, switches, and fixtures for signs of damage, wear and tear, or improper installation.
Test individual components and connections systematically to isolate the source of the problem and determine the appropriate course of action for repairs.

By familiarizing yourself with the basics of wiring, circuitry, and electrical troubleshooting, you can better understand how your home's electrical system works and effectively address common electrical problems as they arise. However, always remember to prioritize safety when working with electricity and consult a licensed electrician if you are unsure or uncomfortable performing electrical repairs yourself.

# Chapter: Guide to Selecting and Using Power Tools Safely and Effectively

In this chapter, we will provide you with a comprehensive guide to selecting and using power tools safely and effectively. Power tools are essential for many home improvement and maintenance projects, but they can be dangerous if not used properly. By following proper safety precautions and techniques, you can minimize the risk of accidents and injuries while achieving professional-quality results.

Selecting the Right Power Tools:

Consider the specific needs of your project and choose power tools that are suitable for the task at hand. Look for tools that are designed for the type of material you will be working with (e.g., wood, metal, concrete).
Research different brands and models to find reliable and durable power tools with features that meet your requirements and budget.
Prioritize safety features such as blade guards, safety switches, and anti-kickback mechanisms when selecting power tools.
Read reviews and ratings from other users to gauge the performance and reliability of different power tool brands and models.
Using Power Tools Safely:

Read the manufacturer's instructions and familiarize yourself with the proper operation and safety precautions for each power tool before use.

Wear appropriate personal protective equipment (PPE), including safety goggles, ear protection, gloves, and dust masks, to protect yourself from hazards such as flying debris, loud noise, and harmful dust.

Ensure that power tools are properly maintained and in good working condition before each use. Inspect cords, switches, blades, and other components for damage or wear and tear, and replace any faulty parts immediately.

Secure workpieces firmly in place using clamps or vises to prevent them from moving or shifting during operation.

Use power tools in a well-ventilated area to minimize exposure to dust, fumes, and other airborne particles.

Consider using dust collection systems or wearing a respirator when working with materials that produce harmful dust or fumes.

Keep hands and fingers away from moving parts and cutting edges of power tools at all times. Never reach over or behind a blade or cutting surface while a power tool is in operation.

Disconnect power tools from electrical outlets or remove batteries when not in use to prevent accidental activation and unauthorized use.

Techniques for Effective Use:

Practice proper body mechanics and posture to reduce strain and fatigue while operating power tools. Maintain a stable stance and keep your body balanced and centered to maintain control and accuracy.

Use the appropriate speed, depth, and pressure settings for each power tool and material to achieve optimal results while minimizing the risk of damage or injury.

Take breaks as needed to rest and recharge, especially during prolonged or repetitive tasks. Fatigue can impair concentration and coordination, increasing the risk of accidents and injuries.

Seek guidance from experienced professionals or attend training workshops to learn proper techniques and best practices for using power tools safely and effectively.

By following these guidelines for selecting and using power tools safely and effectively, you can minimize the risk of accidents and injuries while achieving professional-quality results in your home improvement and maintenance projects. Remember to prioritize safety at all times and consult a licensed professional if you are unsure or uncomfortable performing certain tasks with power tools.

# Chapter: DIY Projects for Home Improvement and Maintenance

In this chapter, we will explore a variety of DIY (do-it-yourself) projects for home improvement and maintenance that utilize power tools and electrical knowledge. These projects are designed to help you enhance the functionality, aesthetics, and efficiency of your home while gaining valuable hands-on experience with power tools and electrical systems.

Installing Ceiling Fans:

Ceiling fans can help improve airflow and circulation in your home, making it more comfortable and energy-efficient. Begin by selecting a suitable location for the ceiling fan and ensuring that it is properly supported by a ceiling joist or support brace.
Use a stud finder to locate ceiling joists and mark the placement of the fan's mounting bracket on the ceiling.
Cut a hole in the ceiling for the electrical box and run electrical wiring to the location of the fan.
Install the fan's mounting bracket and electrical box according to the manufacturer's instructions, and then mount the fan motor and blades.
Connect the electrical wires from the fan to the corresponding wires in the ceiling, ensuring proper grounding and polarity.
Test the fan to ensure it operates correctly and adjust the fan's speed and direction as needed.
Building a Workbench:

A sturdy workbench is essential for DIY enthusiasts and homeowners who enjoy woodworking, crafting, or home repairs.

Start by selecting a suitable location for the workbench and determining the desired dimensions based on available space and intended use.

Cut lumber to size for the workbench frame and assemble it using screws or nails, ensuring that the frame is square and level.

Attach a plywood or hardwood top to the frame using screws, and reinforce the corners and edges with additional bracing if necessary.

Install a vice or clamps to the workbench top for securing workpieces during projects, and add storage shelves or drawers underneath for storing tools and supplies.

Sand the workbench surface smooth and apply a protective finish, such as polyurethane or varnish, to protect against scratches and stains.

Upgrading Kitchen Cabinets:

Updating kitchen cabinets can transform the look and feel of your kitchen without the need for a full remodel.

Begin by removing cabinet doors and hardware, and then sand the surfaces to remove any existing finish or imperfections.

Apply a fresh coat of paint or stain to the cabinets and allow them to dry completely before reattaching doors and hardware.

Consider replacing outdated cabinet hardware with new knobs, handles, or pulls for a modern and cohesive look.

Add organizational accessories such as pull-out shelves, lazy Susans, or drawer dividers to maximize storage and functionality.

Install under-cabinet lighting or task lighting to brighten workspaces and enhance visibility while cooking or preparing meals.

Creating Custom Shelving:

Custom shelving can provide additional storage and display space in any room of your home, from the garage to the living room.
Determine the desired size and configuration of the shelves based on available space and storage needs.
Cut lumber or plywood to size for the shelves and support brackets, and then assemble the components using screws or nails.

Attach the shelves to the wall using brackets or cleats, ensuring they are level and properly anchored to wall studs.
Consider adding decorative trim or molding to the front edges of the shelves for a polished and professional look.
Customize the shelves with paint, stain, or decorative finishes to match your existing decor and personal style.

By tackling these DIY projects for home improvement and maintenance, you can enhance the functionality, aesthetics, and efficiency of your home while gaining valuable hands-on experience with power tools and electrical systems. Remember to prioritize safety at all times and consult a licensed professional if you are unsure or uncomfortable performing certain tasks with power tools or electrical work.

# Chapter: Power Systems and Energy Independence

Introduction to Renewable Energy Sources

In this chapter, we will explore the fascinating world of power systems and renewable energy sources, including solar, wind, and hydro power. As the demand for energy continues to rise and concerns about climate change and environmental sustainability grow, renewable energy has emerged as a viable and increasingly popular alternative to traditional fossil fuels. By harnessing the natural power of the sun, wind, and water, individuals and communities can achieve greater energy independence while reducing their carbon footprint and environmental impact.

Solar Power:

Solar power is generated by capturing sunlight and converting it into electricity using photovoltaic (PV) solar panels or solar thermal collectors.
PV solar panels are composed of silicon cells that produce electricity when exposed to sunlight. These panels can be installed on rooftops or in solar farms to generate clean, renewable energy.
Solar thermal collectors use mirrors or lenses to concentrate sunlight onto a receiver, which heats a fluid to produce steam that drives a turbine and generates electricity.

Solar power is abundant, renewable, and emissions-free, making it an attractive option for homeowners, businesses, and utilities looking to reduce their reliance on fossil fuels and lower their energy costs.

Wind Power:

Wind power is generated by harnessing the kinetic energy of wind and converting it into mechanical or electrical energy using wind turbines.

Wind turbines consist of rotor blades mounted on a tower, which spin when exposed to wind. The rotational energy of the blades is converted into electricity by a generator housed inside the turbine.

Wind power is a mature and rapidly growing renewable energy technology, with wind farms producing electricity on a utility scale and smaller-scale wind turbines installed in rural and remote areas.

Wind power is clean, abundant, and cost-effective, making it a valuable addition to the energy mix and a key component of efforts to combat climate change and transition to a sustainable energy future.

Hydro Power:

Hydro power, also known as hydropower, is generated by harnessing the energy of flowing water, such as rivers, streams, and waterfalls.

Hydroelectric power plants use dams or other structures to capture and store water, which is then released through turbines to generate electricity.

Hydro power is a reliable and proven renewable energy source, with hydroelectric plants supplying a significant portion of the world's electricity.

In addition to traditional hydroelectric dams, newer technologies such as run-of-river hydroelectric systems and tidal energy converters are being developed to tap into the power of flowing water in more sustainable and environmentally friendly ways.

By embracing renewable energy sources such as solar, wind, and hydro power, individuals and communities can achieve greater energy independence, reduce their carbon footprint, and contribute to a cleaner, greener future for generations to come. In the following sections, we will delve deeper into each of these renewable energy technologies, exploring their benefits, challenges, and potential applications in homes, businesses, and communities.

# Chapter: Setting Up and Maintaining Home-Based Renewable Energy Systems

In this chapter, we will provide you with a comprehensive guide to setting up and maintaining home-based renewable energy systems. From solar panels to wind turbines and hydroelectric generators, harnessing renewable energy sources at home can help you reduce your reliance on fossil fuels, lower your energy bills, and contribute to a cleaner, greener planet. We'll cover everything you need to know to get started, from planning and installation to maintenance and troubleshooting.

Assessing Your Energy Needs:

Begin by assessing your current energy usage and identifying areas where renewable energy systems can help meet your needs.
Consider factors such as your average monthly energy consumption, available space for installing renewable energy systems, and local regulations and incentives for renewable energy.
Choosing the Right Renewable Energy System:

Research different renewable energy technologies, including solar, wind, and hydro power, to determine which option is best suited to your location, energy needs, and budget. Consider factors such as the availability of sunlight, wind speed, water flow, and potential obstacles or limitations on your property.

Planning and Installation:

Consult with a qualified renewable energy installer or engineer to assess your property, conduct site surveys, and develop a customized system design that meets your energy needs and budget.
Obtain any necessary permits or approvals from local authorities before beginning installation, and ensure that your chosen system complies with building codes and regulations.
Work with your installer to select high-quality components, such as solar panels, inverters, batteries, and mounting hardware, and ensure that they are properly installed and connected according to manufacturer specifications.
Maintenance and Monitoring:

Regular maintenance is essential for keeping your renewable energy system operating efficiently and maximizing its lifespan.
Perform routine inspections of your system, including checking for signs of damage, corrosion, or wear and tear on components such as solar panels, wind turbines, and batteries.
Clean solar panels regularly to remove dust, dirt, and debris that can reduce their efficiency, and trim any vegetation that may shade the panels and reduce their output.
Monitor your system's performance using data logging or monitoring software, and be alert to any changes or abnormalities that may indicate a problem.
Troubleshooting and Repairs:

In the event of a system malfunction or failure, it's important to diagnose and address the problem as quickly as possible to minimize downtime and ensure continued operation.
Work with your installer or a qualified technician to troubleshoot and repair any issues with your renewable energy system, including electrical faults, mechanical failures, or component malfunctions.

Keep spare parts and tools on hand for routine maintenance and emergency repairs, and maintain a record of your system's performance and maintenance history for future reference.

By following these guidelines for setting up and maintaining home-based renewable energy systems, you can enjoy the benefits of clean, sustainable energy while minimizing your environmental impact and reducing your dependence on fossil fuels. Remember to prioritize safety at all times and consult with qualified professionals for assistance with planning, installation, and maintenance as needed.

# Chapter: Strategies for Achieving Energy Independence

In this chapter, we will explore various strategies for reducing reliance on traditional power grids and achieving energy independence. Whether you're looking to lower your energy bills, reduce your carbon footprint, or prepare for emergencies, there are numerous options available for generating, storing, and managing your own energy supply. From renewable energy sources to energy-efficient technologies and lifestyle changes, these strategies can help you take control of your energy usage and create a more sustainable and resilient home environment.

Embrace Renewable Energy Sources:

Invest in solar panels to generate electricity from the sun's abundant energy. Solar photovoltaic (PV) systems can be installed on rooftops or ground-mounted arrays to capture sunlight and convert it into electricity for use in your home. Consider installing wind turbines to harness the power of the wind and generate clean, renewable electricity. Small-scale wind turbines can be installed on your property to supplement your energy needs and reduce your reliance on the grid.
Explore the potential for hydroelectric power by installing micro-hydro systems or small-scale hydroelectric generators on streams or rivers located on your property. These systems can generate electricity from flowing water and provide a reliable source of renewable energy.
Invest in Energy Storage Solutions:

Install battery storage systems to store excess energy generated by your renewable energy systems for use during periods of low production or high demand. Battery storage systems can help you maximize the value of your renewable energy investments and increase your energy independence. Consider other energy storage options such as thermal storage systems, compressed air energy storage, or flywheel energy storage, depending on your specific needs and circumstances.
Implement Energy-Efficient Technologies and Practices:

Upgrade to energy-efficient appliances, lighting, and HVAC systems to reduce your energy consumption and lower your energy bills. Choose appliances and equipment with high Energy Star ratings and consider smart home technologies that allow you to monitor and control energy usage remotely. Improve the insulation and air sealing of your home to reduce heat loss in the winter and heat gain in the summer. Seal gaps and cracks in windows, doors, and walls, and add insulation to attics, basements, and crawl spaces to improve energy efficiency and comfort.
Practice energy-saving habits such as turning off lights and appliances when not in use, adjusting thermostats to conserve energy, and using natural ventilation and daylighting to reduce reliance on artificial lighting and HVAC systems.
Explore Off-Grid Living and Self-Sufficiency:

Consider transitioning to off-grid living by disconnecting from the traditional power grid and relying solely on renewable energy sources and energy storage solutions for your electricity needs.
Implement self-sufficiency measures such as rainwater harvesting, greywater recycling, and sustainable food production to reduce dependence on municipal utilities and external resources.

Invest in backup power systems such as generators or fuel cells to provide additional energy security and resilience during power outages or emergencies.

By implementing these strategies for reducing reliance on traditional power grids and achieving energy independence, you can create a more sustainable, resilient, and self-sufficient home environment. Whether you're motivated by environmental concerns, economic benefits, or the desire for greater energy security, there are numerous options available for taking control of your energy future and reducing your dependence on fossil fuels.

# Chapter: Successful Off-Grid Living and Sustainable Energy Projects

In this chapter, we will explore inspiring examples of successful off-grid living and sustainable energy projects from around the world. These innovative initiatives demonstrate the feasibility and benefits of living off-grid while harnessing renewable energy sources to power homes, communities, and businesses. From remote cabins and eco-villages to sustainable farms and off-grid retreats, these projects showcase the ingenuity, resilience, and dedication of individuals and communities striving to live in harmony with nature while reducing their environmental impact.

Off-Grid Eco-Village: Dancing Rabbit Ecovillage, Missouri, USA

Dancing Rabbit Ecovillage is a thriving intentional community located in rural Missouri, dedicated to sustainable living and environmental stewardship.
The community consists of ecologically designed homes, gardens, and communal spaces powered by renewable energy sources such as solar panels, wind turbines, and biomass generators.
Residents of Dancing Rabbit embrace a low-impact lifestyle, practicing permaculture, organic farming, and resource conservation to minimize their ecological footprint and live in harmony with nature.
Off-Grid Sustainable Farm: Finca Luna Nueva, Costa Rica

Finca Luna Nueva is a certified organic farm and eco-lodge located in the rainforests of Costa Rica, dedicated to regenerative agriculture and sustainable living practices.

The farm operates entirely off-grid, generating electricity from solar panels and hydropower systems while utilizing biodigesters and composting toilets to manage waste and produce renewable energy.

Guests can participate in farm-to-table experiences, wellness retreats, and educational workshops focused on sustainable agriculture, permaculture, and tropical ecology.

Off-Grid Retreat Center: Lost Valley Education and Event Center, Oregon, USA

Lost Valley is an off-grid education and event center located in the Pacific Northwest, offering workshops, retreats, and community gatherings focused on sustainability, holistic living, and personal growth.

The center operates off-grid, relying on solar panels, wind turbines, and micro-hydro systems to generate electricity, while rainwater harvesting and greywater recycling systems provide water for drinking, irrigation, and sanitation.

Visitors to Lost Valley can experience sustainable living firsthand through hands-on workshops, educational tours, and immersive retreats focused on renewable energy, natural building, and ecological design.

Off-Grid Tiny House Community: Earthship Biotecture, New Mexico, USA

Earthship Biotecture is an off-grid sustainable housing community located in the high desert of New Mexico, dedicated to promoting self-sufficiency, resilience, and environmental responsibility.

The community consists of Earthship homes—off-grid, passive solar structures built from recycled materials such as tires, bottles, and cans—that utilize solar power, rainwater harvesting, and greywater recycling systems to achieve self-sufficiency.

Earthship residents embrace a simple, sustainable lifestyle, growing their own food, harvesting rainwater, and generating renewable energy to meet their basic needs while minimizing their impact on the environment.

These successful off-grid living and sustainable energy projects serve as inspiring examples of what is possible when individuals and communities come together to prioritize environmental stewardship, self-sufficiency, and resilience. By embracing renewable energy sources, adopting sustainable living practices, and fostering a sense of community and collaboration, we can create a more sustainable and resilient future for ourselves and generations to come.

# Chapter: Survival and Emergency Preparedness

Importance of Being Prepared for Emergencies and Disasters

In this chapter, we will discuss the importance of survival and emergency preparedness and why being ready for unexpected events is crucial for safeguarding yourself, your loved ones, and your community. From natural disasters to man-made crises, emergencies can strike at any time and in any place, and being prepared can mean the difference between life and death.

Understanding the Risks:

Natural disasters such as hurricanes, earthquakes, floods, wildfires, and tornadoes can cause widespread devastation and displacement, disrupting infrastructure, communication networks, and access to essential services.
Man-made disasters such as industrial accidents, terrorist attacks, and pandemics can also pose significant threats to public safety and security, requiring swift and coordinated emergency response efforts to mitigate their impact.
Benefits of Emergency Preparedness:

By being prepared for emergencies, you can minimize the risks to yourself and your loved ones, reduce the severity of injuries and property damage, and increase the likelihood of survival and recovery.

Preparedness measures such as creating emergency plans, assembling survival kits, and practicing evacuation drills can help you respond effectively to emergencies, stay calm under pressure, and make informed decisions in stressful situations. Being prepared for emergencies can also empower you to help others in need, whether it's providing first aid, offering shelter and support, or coordinating community response efforts.

Key Components of Emergency Preparedness:

Develop an Emergency Plan: Create a comprehensive emergency plan that outlines evacuation routes, communication protocols, emergency contacts, and essential supplies and resources. Review and practice your plan regularly with family members and neighbors to ensure everyone knows what to do in an emergency.

Assemble a Survival Kit: Prepare a well-stocked survival kit containing essential items such as water, non-perishable food, first aid supplies, medications, flashlights, batteries, blankets, and personal hygiene products. Store your kit in a secure, easily accessible location and periodically check and replenish its contents as needed.

Stay Informed: Stay informed about potential hazards and emergencies in your area by monitoring weather forecasts, emergency alerts, and news updates from reliable sources. Sign up for emergency notification systems and community alerts to receive timely information and instructions during emergencies.

Build Community Resilience: Get involved in community preparedness and resilience-building efforts by joining local emergency response teams, volunteering with disaster relief organizations, and participating in community emergency planning and training activities.

Taking Action:

Take proactive steps to mitigate the risks of emergencies and disasters by implementing preventive measures such as securing your home, maintaining emergency supplies, and reinforcing critical infrastructure.

Stay vigilant and prepared for emergencies at all times, regardless of the season or perceived level of risk. Be ready to adapt and respond quickly to changing circumstances and emerging threats.

Encourage others to prioritize emergency preparedness and resilience-building efforts by sharing information, resources, and best practices with friends, family members, and community members.

By taking proactive steps to prepare for emergencies and disasters, you can increase your resilience, reduce the impact of emergencies, and protect yourself and your loved ones from harm. Remember, being prepared is not just about having supplies and plans in place—it's about developing a mindset of readiness, resilience, and adaptability that can help you navigate and overcome any challenge that comes your way.

# Chapter: Creating a Comprehensive Emergency Preparedness Plan for Your Home and Family

In this chapter, we will guide you through the process of creating a comprehensive emergency preparedness plan for your home and family. By developing a well-thought-out plan and taking proactive steps to prepare for emergencies, you can minimize risks, enhance safety, and ensure the well-being of yourself and your loved ones in times of crisis.

Assessing Risks and Hazards:

Begin by identifying potential risks and hazards that may affect your area, such as natural disasters (e.g., earthquakes, hurricanes, wildfires), man-made emergencies (e.g., chemical spills, terrorist attacks), and health emergencies (e.g., pandemics, medical emergencies).
Research historical data, local emergency plans, and hazard maps to gain a better understanding of the types of emergencies that are most likely to occur in your region and their potential impact on your home and community.
Developing an Emergency Plan:

Establish clear communication channels and protocols for staying in touch with family members during emergencies. Designate an out-of-area contact person who can serve as a central point of contact for coordinating communication and relaying information.

Create a family emergency plan that outlines evacuation routes, assembly points, emergency contacts, and designated meeting locations both inside and outside your home. Ensure that all family members are familiar with the plan and know what to do in various emergency scenarios.

Customize your emergency plan to address the specific needs and requirements of family members with disabilities, medical conditions, or special considerations. Include provisions for pets, livestock, and other animals in your plan as well.

Assembling Emergency Supplies:

Assemble a well-stocked emergency supply kit containing essential items to sustain your family for at least 72 hours during an emergency. Include items such as water, non-perishable food, first aid supplies, medications, flashlights, batteries, blankets, clothing, and personal hygiene products.

Customize your emergency kit to meet the specific needs of your family, such as infant formula, diapers, pet food, or prescription medications. Consider including items for comfort and entertainment, such as books, games, and activities for children.

Store your emergency supplies in a durable, waterproof container or backpack that is easy to transport and access in case of evacuation. Keep your kit in a designated location that is easily accessible to all family members and ensure that everyone knows where it is located.

Practicing and Revising Your Plan:

Regularly review and practice your emergency plan with family members to ensure that everyone knows their roles and responsibilities and understands what to do in an emergency. Conduct emergency drills and simulations to simulate different scenarios and test your family's readiness and response capabilities. Use these exercises as opportunities to identify areas for improvement and make adjustments to your plan as needed.

Stay informed about potential hazards and updates to your local emergency plans and advisories. Keep emergency contact information and important documents up to date, and periodically review and revise your emergency plan to reflect changes in your circumstances or environment.

By following these steps and creating a comprehensive emergency preparedness plan for your home and family, you can enhance safety, resilience, and peace of mind in the face of emergencies and disasters. Remember, being prepared is not just about having supplies and plans in place — it's about developing a mindset of readiness, adaptability, and community support that can help you navigate and overcome any challenge that comes your way.

# Chapter: Essential Survival Skills

Including First Aid, Fire Starting, Shelter Building, and Foraging

In this chapter, we will cover essential survival skills that are crucial for thriving in emergency situations and outdoor environments. Whether you're faced with a natural disaster, a wilderness adventure gone awry, or an unexpected emergency, having a solid foundation of survival skills can mean the difference between life and death. We'll explore key techniques for first aid, fire starting, shelter building, and foraging to help you stay safe and secure in any situation.

First Aid:

First aid skills are essential for providing immediate care and assistance to individuals who are injured or experiencing medical emergencies.
Learn basic first aid techniques such as assessing the scene for safety, checking for responsiveness, and performing CPR (cardiopulmonary resuscitation) if necessary.
Familiarize yourself with techniques for controlling bleeding, treating wounds, splinting fractures, and managing common medical emergencies such as burns, bites, and allergic reactions.
Carry a first aid kit with essential supplies such as bandages, gauze pads, antiseptic wipes, adhesive tape, scissors, tweezers, and over-the-counter medications for pain relief and symptom management.
Fire Starting:

Fire is essential for warmth, cooking, signaling, and morale in survival situations. Learning how to start a fire using various methods can be a lifesaving skill.
Practice traditional fire-starting techniques such as friction fire methods (e.g., bow drill, hand drill) and flint and steel.
Carry fire-starting tools in your survival kit, such as waterproof matches, lighters, fire starters, and magnesium fire starters.
Gather and prepare tinder, kindling, and fuel before attempting to start a fire, and ensure that your fire is safely contained and properly extinguished when not in use.
Shelter Building:

Shelter provides protection from the elements and helps regulate body temperature in survival situations. Knowing how to build shelters using natural materials is essential for staying safe and comfortable outdoors.
Learn basic shelter designs such as lean-tos, debris shelters, and A-frame shelters, and practice building them using branches, leaves, grass, and other available materials.
Choose a shelter location that is dry, sheltered from wind and precipitation, and close to a water source if possible.
Insulate your shelter with additional layers of natural materials such as leaves, grass, or pine needles to improve warmth and comfort.
Foraging:

Foraging for wild edible plants and other resources can supplement your food and water supplies in survival situations.
Learn to identify common edible plants, fruits, nuts, and mushrooms in your area, as well as potentially toxic or harmful species to avoid.
Use guidebooks, online resources, and local experts to expand your knowledge of wild edibles and foraging techniques.

Practice sustainable foraging practices such as harvesting only what you need, avoiding rare or endangered species, and respecting wildlife habitats and ecosystems.

By mastering these essential survival skills—including first aid, fire starting, shelter building, and foraging—you can increase your confidence, resilience, and self-reliance in emergency situations and outdoor environments. Remember to practice these skills regularly, seek additional training and guidance as needed, and prioritize safety at all times. With preparation, knowledge, and resourcefulness, you can overcome adversity and thrive in any survival scenario.

# Chapter: Tips for Stockpiling Emergency Supplies and Creating a Survival Kit

In this chapter, we will discuss essential tips for stockpiling emergency supplies and creating a comprehensive survival kit. Whether you're preparing for natural disasters, outdoor adventures, or unexpected emergencies, having a well-stocked survival kit can provide you with the tools and resources you need to stay safe, secure, and self-sufficient in any situation. We'll cover key considerations for selecting and storing emergency supplies, as well as guidelines for assembling a survival kit that meets your specific needs and circumstances.

Assess Your Needs:

Begin by assessing your individual, family, and household needs to determine the types and quantities of emergency supplies you will need to stockpile.
Consider factors such as the size of your household, any specific medical conditions or dietary restrictions, and the types of emergencies that are most likely to occur in your area.
Prioritize Essentials:

Focus on acquiring essential items that are critical for survival and well-being in emergency situations, such as water, food, shelter, first aid supplies, and tools for communication and self-defense.
Start with the basics, such as water, non-perishable food, and first aid supplies, and then gradually expand your stockpile to include additional items based on your needs and priorities.

Follow the Rule of Threes:

Use the "Rule of Threes" as a guideline for prioritizing emergency supplies: you can survive for approximately three minutes without air, three hours without shelter, three days without water, and three weeks without food.
Focus on securing supplies that address these fundamental needs first, and then consider additional items for comfort, hygiene, and longer-term survival.
Consider Storage and Shelf Life:

Choose durable, waterproof containers for storing emergency supplies, such as plastic bins, dry bags, or backpacks with waterproof liners.
Rotate perishable items such as food and medications regularly to ensure freshness and safety, and check expiration dates on non-perishable items periodically to ensure they remain viable.
Store your survival kit in a cool, dry, and easily accessible location that is free from pests, moisture, and extreme temperatures.
Customize Your Kit:

Tailor your survival kit to meet your specific needs and circumstances, considering factors such as climate, terrain, and the types of emergencies that are most likely to occur in your area.
Include personal items such as medications, prescription glasses, copies of important documents, and comfort items such as blankets, clothing, and hygiene products.
Consider the unique needs of family members, including infants, children, elderly individuals, and pets, and include items such as diapers, formula, medications, and pet supplies as needed.
Periodically Review and Update:

Regularly review and update your survival kit to ensure that it remains current, relevant, and effective for your needs. Conduct inventory checks and replace expired or depleted items as needed, and make adjustments to your kit based on changes in your circumstances or environment.

Stay informed about emerging threats, new technologies, and best practices for emergency preparedness, and incorporate this information into your survival planning and stockpiling efforts.

By following these tips for stockpiling emergency supplies and creating a survival kit, you can increase your resilience, readiness, and self-sufficiency in the face of emergencies and disasters. Remember to prioritize essentials, customize your kit to meet your specific needs, and regularly review and update your supplies to ensure that you are well-prepared for whatever challenges may arise. With preparation, foresight, and resourcefulness, you can navigate emergencies with confidence and peace of mind.

# Chapter: Handling Natural Disasters

Understanding Different Types of Natural Disasters and Their Potential Impact

In this chapter, we will explore the various types of natural disasters that can occur and discuss their potential impact on communities, infrastructure, and individuals. From hurricanes and earthquakes to floods and wildfires, natural disasters can pose significant threats to public safety, property, and well-being. By understanding the characteristics and effects of different types of natural disasters, individuals and communities can better prepare for and mitigate their impact, increase resilience, and enhance safety and survival.

Hurricanes and Tropical Storms:

Hurricanes and tropical storms are powerful cyclonic storms characterized by strong winds, heavy rainfall, storm surges, and flooding.
These storms can cause extensive damage to coastal areas, infrastructure, and property, as well as disrupt transportation, communication, and utilities.
Key preparedness measures for hurricanes and tropical storms include evacuating from vulnerable areas, securing property, stocking up on emergency supplies, and staying informed about weather forecasts and evacuation orders.
Earthquakes:

Earthquakes are sudden movements of the earth's crust caused by the release of accumulated stress along fault lines. These seismic events can result in shaking, ground rupture, landslides, and tsunamis, causing widespread damage to buildings, roads, bridges, and infrastructure.
Preparedness measures for earthquakes include securing furniture and heavy objects, reinforcing buildings and structures, creating evacuation plans, and practicing "drop, cover, and hold on" drills.
Floods:

Floods occur when water overflows onto normally dry land, inundating homes, businesses, and infrastructure.
Floods can result from heavy rainfall, snowmelt, storm surges, dam failures, or urban runoff, and can cause extensive damage to property, crops, and infrastructure.
Preparedness measures for floods include elevating buildings, installing flood barriers and sandbags, creating evacuation plans, and purchasing flood insurance to protect against financial losses.
Wildfires:

Wildfires are uncontrolled fires that spread rapidly through vegetation, forests, and grasslands, fueled by dry conditions, high winds, and flammable materials.
These fires can destroy homes, forests, and wildlife habitats, as well as threaten lives, property, and air quality.
Preparedness measures for wildfires include creating defensible space around homes, removing flammable vegetation, maintaining fire-resistant landscaping, and developing evacuation plans.
Tornadoes:

Tornadoes are violent rotating columns of air that extend from thunderstorms to the ground, causing destruction along their path.

These storms can produce strong winds, hail, and flying debris, resulting in damage to buildings, vehicles, and infrastructure.

Preparedness measures for tornadoes include seeking shelter in a sturdy building or underground shelter, staying away from windows and exterior walls, and monitoring weather alerts and warnings.

Landslides and Mudslides:

Landslides and mudslides are mass movements of rock, soil, and debris down a slope, often triggered by heavy rainfall, earthquakes, or volcanic activity.

These events can bury homes, roads, and infrastructure, as well as disrupt transportation and utilities.

Preparedness measures for landslides and mudslides include avoiding steep slopes and areas prone to erosion, monitoring weather and ground conditions, and evacuating from areas at risk.

By understanding the characteristics and effects of different types of natural disasters and implementing appropriate preparedness measures, individuals and communities can reduce their vulnerability, increase resilience, and enhance safety and survival in the face of emergencies and disasters. Remember to stay informed about potential hazards, develop emergency plans, and take proactive steps to protect yourself, your loved ones, and your property from harm. With preparation, foresight, and resourcefulness, you can navigate natural disasters with confidence and resilience.

# Chapter: Preparing for Earthquakes, Hurricanes, Wildfires, Floods, and Other Natural Calamities

In this chapter, we will discuss essential steps for preparing for earthquakes, hurricanes, wildfires, floods, and other natural calamities. Each type of natural disaster presents unique challenges and requires specific preparedness measures to ensure the safety and well-being of individuals, families, and communities. By understanding the risks and taking proactive steps to prepare for these events, you can increase resilience, minimize damage, and enhance survival in the face of emergencies and disasters.

Earthquakes:

Secure heavy furniture, appliances, and objects to prevent them from falling or causing injury during an earthquake. Identify safe locations in your home or workplace where you can take cover during an earthquake, such as under sturdy tables or desks, and practice "drop, cover, and hold on" drills regularly.
Retrofit older buildings and structures to improve their earthquake resistance, and consider purchasing earthquake insurance to protect against financial losses.
Hurricanes:

Know your evacuation zone and develop a hurricane evacuation plan that includes multiple evacuation routes, designated shelters, and emergency contacts.

Stock up on essential supplies such as water, non-perishable food, medications, batteries, flashlights, and first aid supplies, and secure your home by installing storm shutters, reinforcing doors and windows, and trimming trees and shrubs.

Stay informed about weather forecasts, evacuation orders, and emergency alerts, and be prepared to evacuate if necessary to avoid dangerous storm surge, flooding, or high winds.

Wildfires:

Create defensible space around your home by clearing vegetation, leaves, and debris from the immediate vicinity and using fire-resistant landscaping materials.

Prepare for potential evacuations by packing a "go bag" with essential items such as clothing, medications, important documents, and pet supplies, and develop an evacuation plan that includes multiple evacuation routes and meeting locations.

Stay informed about wildfire risks, fire weather conditions, and evacuation orders by monitoring local news, weather alerts, and emergency notifications.

Floods:

Know your flood risk and elevation level, and consider purchasing flood insurance to protect against financial losses from flood damage.

Elevate utilities, appliances, and electrical systems above potential flood levels, and install flood barriers, sandbags, or flood-proofing measures to prevent water intrusion into your home.

Create an emergency flood plan that includes evacuation routes, emergency contacts, and designated meeting locations, and stay informed about flood warnings and evacuation orders.

Other Natural Calamities:

Prepare for other natural calamities such as tornadoes, landslides, tsunamis, and volcanic eruptions by familiarizing yourself with local hazards, evacuation routes, and emergency procedures.

Stockpile essential supplies and emergency equipment, develop communication and reunification plans with family members, neighbors, and coworkers, and practice emergency drills and evacuation procedures regularly.
Stay informed about potential hazards and emergency situations by monitoring weather forecasts, seismic activity, and other relevant information sources, and be prepared to take swift and decisive action to protect yourself and your loved ones.

By taking proactive steps to prepare for earthquakes, hurricanes, wildfires, floods, and other natural calamities, you can increase resilience, enhance safety, and improve survival outcomes in the face of emergencies and disasters. Remember to stay informed, stay prepared, and stay vigilant, and prioritize the safety and well-being of yourself, your loved ones, and your community at all times. With preparation, foresight, and resourcefulness, you can navigate natural calamities with confidence and resilience.

# Chapter: Evacuation Planning, Sheltering in Place, and Community Response Strategies

In this chapter, we will discuss essential strategies for evacuation planning, sheltering in place, and coordinating community response efforts during emergencies and disasters. Whether faced with natural calamities, hazardous incidents, or other emergency situations, having a well-developed plan, and understanding community response protocols can help individuals and communities stay safe, secure, and resilient in times of crisis. We will explore key considerations for evacuating safely, sheltering in place effectively, and fostering community cooperation and support in emergency situations.

Evacuation Planning:

Identify potential evacuation routes, assembly points, and designated shelters in your community, and develop a comprehensive evacuation plan for your household or organization.
Determine triggers and thresholds for evacuation orders, such as severe weather warnings, hazardous material spills, or other imminent threats, and establish communication protocols for disseminating evacuation alerts and instructions.

Pack emergency supplies and essential items in advance, including water, food, medications, clothing, important documents, and pet supplies, and store them in a portable "go bag" or vehicle emergency kit for quick and easy access during evacuations.

Practice evacuation drills regularly with family members, employees, or community members to familiarize everyone with evacuation routes, procedures, and safety protocols, and ensure that everyone knows their roles and responsibilities during evacuations.

Sheltering in Place:

In some emergency situations, it may be safer to shelter in place rather than evacuate, particularly if evacuation routes are blocked or conditions are too hazardous to travel.

Identify a secure location within your home, workplace, or community where you can shelter in place safely, such as a basement, storm cellar, or interior room with no windows.

Stockpile emergency supplies and essential items for sheltering in place, including water, non-perishable food, first aid supplies, flashlights, batteries, and communication devices, and ensure that you have adequate ventilation, sanitation, and protection from hazards such as airborne contaminants or extreme temperatures.

Follow guidance from local authorities and emergency management agencies regarding shelter-in-place orders, evacuation orders, and other protective actions, and stay informed about changing conditions and updates through official channels and communication platforms.

Community Response Strategies:

Foster collaboration and coordination among community members, organizations, and agencies to facilitate an effective and cohesive response to emergencies and disasters.

Establish community emergency response teams, neighborhood watch groups, or volunteer networks to support emergency responders, provide assistance to vulnerable populations, and disseminate critical information and resources during emergencies.

Develop partnerships with local government agencies, emergency services, nonprofit organizations, businesses, and community groups to enhance preparedness, response, and recovery efforts, and leverage existing resources and expertise to address community needs and priorities.

Engage in community-wide training, exercises, and preparedness initiatives to build awareness, resilience, and capacity for responding to emergencies and disasters, and empower individuals and communities to take proactive steps to protect themselves and their neighbors.

By implementing effective evacuation planning, sheltering in place strategies, and community response measures, individuals and communities can enhance safety, resilience, and survival in the face of emergencies and disasters.

Remember to stay informed, stay prepared, and stay connected with your neighbors and community partners, and prioritize the safety and well-being of yourself, your loved ones, and your community at all times. With preparation, collaboration, and resilience, we can navigate emergencies and disasters with confidence and compassion, and emerge stronger and more resilient than before.

# Chapter: Post-Disaster Recovery and Rebuilding Efforts

In this chapter, we will explore the critical process of post-disaster recovery and rebuilding efforts following emergencies and disasters. When communities are impacted by natural calamities, hazardous incidents, or other disruptive events, the journey towards recovery can be long and challenging. However, with resilience, determination, and community cooperation, affected areas can rebuild and thrive once again. We will discuss key components of post-disaster recovery, including assessing damage, mobilizing resources, fostering community resilience, and rebuilding infrastructure and livelihoods.

Assessing Damage and Needs:

Immediately following a disaster, conduct comprehensive damage assessments to evaluate the extent of destruction to homes, buildings, infrastructure, and natural resources. Identify priority areas and critical needs, such as emergency shelter, food, water, medical care, sanitation, and safety, and prioritize resources and assistance accordingly.
Engage with affected communities, individuals, and stakeholders to gather input, assess needs, and develop recovery plans that are responsive to local priorities and circumstances.

Mobilizing Resources and Assistance:

Coordinate with government agencies, non-profit organizations, businesses, and international partners to mobilize resources, supplies, and assistance for affected communities.

Provide financial assistance, grants, loans, and other forms of support to individuals, businesses, and organizations impacted by the disaster, and ensure equitable access to resources for vulnerable populations and marginalized communities.

Establish disaster recovery centers, resource distribution points, and information hubs to facilitate access to assistance, information, and services for affected individuals and families.

Fostering Community Resilience:

Build community resilience by promoting social cohesion, mutual aid, and collective action among residents, organizations, and stakeholders.

Provide psychosocial support, counseling, and mental health services to individuals and families affected by the disaster, and facilitate peer support networks and community healing activities.

Engage in community-led initiatives, capacity-building activities, and participatory decision-making processes to empower residents and stakeholders to contribute to the recovery and rebuilding efforts.

Rebuilding Infrastructure and Livelihoods:

Invest in rebuilding and repairing critical infrastructure such as roads, bridges, utilities, schools, hospitals, and housing to restore essential services and support economic recovery.

Promote sustainable and resilient rebuilding practices that enhance disaster resistance, environmental sustainability, and community well-being.

Support local businesses, industries, and livelihoods by providing financial assistance, technical support, and economic development initiatives to stimulate growth and create employment opportunities.

Learning and Adaptation:

Capture lessons learned from the disaster and recovery process to inform future preparedness, mitigation, and response efforts.
Conduct post-disaster evaluations, reviews, and assessments to identify strengths, weaknesses, gaps, and opportunities for improvement in disaster management and recovery systems.
Share best practices, innovative approaches, and success stories from the recovery process to inspire and inform other communities facing similar challenges.

By prioritizing post-disaster recovery and rebuilding efforts, communities can recover from adversity, restore normalcy, and emerge stronger and more resilient than before. Remember to leverage local knowledge, resources, and capacities, and prioritize the needs and aspirations of affected individuals and communities throughout the recovery process. With collaboration, innovation, and perseverance, we can build back better and create a more resilient and sustainable future for all.

# Chapter: Water Supply and Management

Importance of Clean Water for Survival and Sustainability

In this chapter, we will delve into the critical importance of clean water for survival and sustainability, as well as strategies for managing water supply effectively in both everyday life and emergency situations. Water is essential for life, supporting hydration, sanitation, agriculture, industry, and ecosystem health. Ensuring access to clean and safe water is paramount for human health, environmental integrity, and community well-being. We will explore the significance of clean water, challenges to water availability and quality, and approaches to sustainable water management.

The Significance of Clean Water:

Clean water is essential for human survival, supporting hydration, digestion, circulation, and overall health and well-being.
Access to clean water is critical for sanitation and hygiene, preventing waterborne diseases, and reducing the spread of illness and infection.
Clean water is also vital for agriculture, providing irrigation for crops, livestock, and food production, as well as supporting ecosystems, biodiversity, and natural habitats.
Challenges to Water Availability and Quality:

Despite its importance, access to clean water remains a challenge for millions of people worldwide, particularly in developing countries and regions affected by poverty, conflict, and environmental degradation.

Water scarcity, pollution, over-extraction, climate change, and inadequate infrastructure pose significant threats to water availability and quality, exacerbating inequalities and jeopardizing human health and environmental sustainability. Contamination of water sources by pollutants, pathogens, chemicals, and industrial waste can have severe consequences for human health, ecosystems, and economic development, leading to waterborne diseases, ecological degradation, and economic losses.

Approaches to Sustainable Water Management:

Adopt integrated water management approaches that balance competing water uses, promote conservation and efficiency, and protect water resources for present and future generations.

Invest in water infrastructure, technology, and innovation to improve access to clean water, treat wastewater, and enhance water quality and safety.

Implement watershed management and ecosystem restoration initiatives to protect and restore natural water systems, including rivers, lakes, wetlands, and aquifers, and promote biodiversity and ecological resilience.

Encourage community engagement, stakeholder collaboration, and participatory decision-making processes to ensure inclusive and equitable water governance, management, and allocation.

Water Supply and Management in Emergency Situations:

In emergency situations such as natural disasters, conflicts, and humanitarian crises, ensuring access to clean water is a top priority for saving lives, preventing disease outbreaks, and supporting relief and recovery efforts.

Establish emergency water supply systems, distribution points, and treatment facilities to provide safe drinking water to affected populations, and prioritize vulnerable groups such as children, elderly individuals, and people with disabilities. Promote water conservation, hygiene promotion, and sanitation practices to prevent waterborne diseases and promote public health and well-being in emergency settings.

By recognizing the significance of clean water for survival and sustainability, and implementing sustainable water management practices, we can safeguard this precious resource for future generations and build resilient and healthy communities. Remember to prioritize water conservation, protection, and equitable access, and advocate for policies and actions that promote water security, human rights, and environmental stewardship. With collective action, innovation, and commitment, we can ensure access to clean water for all and create a more sustainable and prosperous future.

# Chapter: Methods for Ensuring a Reliable and Safe Water Supply at Home

Ensuring access to a reliable and safe water supply at home is essential for maintaining health, hygiene, and overall well-being. In this chapter, we will explore various methods and technologies for securing clean and safe water for household use. From filtration and purification systems to conservation practices and maintenance strategies, there are several approaches that homeowners can take to ensure a consistent and high-quality water supply. Let's delve into some of these methods:

Water Filtration Systems:

Install a point-of-use water filtration system to remove impurities, contaminants, and odors from tap water. Common types of filtration systems include activated carbon filters, reverse osmosis systems, and sediment filters.
Choose a filtration system that is certified by reputable organizations such as NSF International or the Water Quality Association to ensure effectiveness and reliability.
Regularly replace filter cartridges according to manufacturer recommendations to maintain optimal performance and water quality.
Water Purification Methods:

Consider additional purification methods such as ultraviolet (UV) disinfection, ozonation, or distillation to further enhance water safety and purity.

UV disinfection systems use ultraviolet light to destroy bacteria, viruses, and other microorganisms in water, providing an extra layer of protection against waterborne diseases.

Ozonation systems inject ozone gas into water to neutralize contaminants and improve taste and odor, while distillation systems heat water to vaporize impurities and then condense the steam into purified water.

Boiling Water:

Boiling water is a simple and effective method for purifying water at home, killing harmful bacteria, parasites, and viruses that may be present.

Bring water to a rolling boil for at least one minute (or three minutes at higher altitudes) to ensure complete disinfection.

Allow boiled water to cool before drinking or using for cooking, and store it in clean, covered containers to prevent recontamination.

Water Conservation Practices:

Practice water conservation at home to reduce water consumption, lower utility bills, and minimize environmental impact.

Fix leaky faucets, toilets, and pipes promptly to prevent water waste and loss.

Install water-efficient fixtures and appliances such as low-flow toilets, showerheads, and aerators to reduce water usage without sacrificing performance.

Regular Maintenance and Testing:

Conduct regular maintenance checks on plumbing fixtures, water heaters, and filtration systems to ensure proper function and performance.

Schedule annual water quality tests to assess the safety and potability of your home's water supply, and address any issues or concerns promptly.

Keep records of maintenance activities and test results to track changes in water quality over time and identify potential problems early.

By implementing these methods for ensuring a reliable and safe water supply at home, homeowners can enjoy peace of mind knowing that their water is clean, safe, and healthy for drinking, cooking, bathing, and other household activities. Remember to prioritize regular maintenance, testing, and monitoring to keep your water supply in optimal condition, and take proactive steps to address any concerns or issues as they arise. With proper care and attention, you can maintain a consistent and high-quality water supply for your home and family for years to come.

# Chapter: Rainwater Harvesting, Water Purification Techniques, and Water Conservation Strategies

In this chapter, we will explore the benefits of rainwater harvesting, various water purification techniques, and effective water conservation strategies. These approaches play a vital role in ensuring access to clean and sustainable water sources, reducing reliance on traditional water supplies, and promoting environmental sustainability. By harnessing rainwater, purifying water effectively, and conserving water wisely, individuals and communities can enhance water security, resilience, and sustainability.

Rainwater Harvesting:

Rainwater harvesting involves collecting and storing rainwater from rooftops, surfaces, or catchment areas for later use.
Install a rainwater harvesting system with gutters, downspouts, and storage tanks to capture rainwater runoff from rooftops and direct it into storage containers.
Choose food-grade, UV-resistant storage tanks or cisterns to store collected rainwater safely and prevent contamination.
Use harvested rainwater for non-potable purposes such as landscape irrigation, gardening, toilet flushing, and laundry to reduce demand on municipal water supplies and conserve freshwater resources.

Water Purification Techniques:

Implement water purification techniques to treat harvested rainwater or other water sources for safe and potable use.
Consider using filtration systems, such as activated carbon filters, ceramic filters, or sediment filters, to remove impurities, sediment, and debris from water.
Explore additional purification methods such as ultraviolet (UV) disinfection, ozonation, or distillation to eliminate bacteria, viruses, and other pathogens from water and improve water quality.
Follow manufacturer instructions and guidelines for operating and maintaining water purification systems to ensure effectiveness and reliability.

Water Conservation Strategies:

Adopt water conservation strategies to reduce water usage, minimize waste, and promote efficient water management practices.
Install water-saving fixtures and appliances such as low-flow toilets, showerheads, faucets, and aerators to reduce water consumption without compromising performance.
Practice simple water-saving habits such as turning off faucets when not in use, fixing leaks promptly, and using a broom instead of a hose to clean outdoor surfaces.
Incorporate water-wise landscaping techniques such as xeriscaping, mulching, and native plantings to reduce outdoor water usage and promote sustainable landscaping practices.

Integrated Water Management:

Integrate rainwater harvesting, water purification, and water conservation practices into holistic water management approaches to maximize efficiency and sustainability.

Design and implement integrated water management systems that combine multiple water sources, treatment methods, and conservation measures to meet diverse water needs while minimizing environmental impact.
Consider factors such as local climate, hydrology, land use, and water demand when planning and implementing integrated water management strategies, and engage stakeholders and communities in decision-making processes to ensure inclusivity and equity.

By combining rainwater harvesting, water purification techniques, and water conservation strategies, individuals, households, and communities can enhance water security, promote environmental sustainability, and build resilience to water-related challenges. Remember to prioritize regular maintenance, monitoring, and education to ensure the effectiveness and sustainability of these approaches, and empower individuals and communities to take proactive steps towards sustainable water management. With creativity, innovation, and commitment, we can harness the power of water wisely and responsibly for the benefit of present and future generations.

# Chapter: Managing Wastewater and Sewage in an Environmentally Responsible Manner

In this chapter, we will explore the importance of managing wastewater and sewage in an environmentally responsible manner. Effective wastewater management is crucial for protecting public health, safeguarding water resources, and preserving environmental quality. By implementing sustainable wastewater treatment and disposal practices, communities can minimize pollution, reduce environmental impact, and promote sustainable development.

Understanding Wastewater and Sewage:

Wastewater refers to any water that has been used in homes, businesses, industries, or agriculture and contains contaminants, pollutants, and organic matter.
Sewage specifically refers to wastewater from toilets, sinks, showers, and other domestic sources that contains human waste and pathogens.

Proper management of wastewater and sewage is essential to prevent the spread of waterborne diseases, protect water quality, and maintain ecological balance in aquatic ecosystems.

Wastewater Treatment Processes:

Wastewater treatment involves a series of physical, chemical, and biological processes to remove contaminants and pollutants from wastewater before it is discharged back into the environment.

Primary treatment involves screening, sedimentation, and filtration to remove solid particles and large debris from wastewater.

Secondary treatment utilizes biological processes such as activated sludge, trickling filters, or constructed wetlands to break down organic matter and remove dissolved pollutants.

Tertiary treatment employs advanced filtration, disinfection, and nutrient removal techniques to further purify wastewater and meet stringent water quality standards.

Sustainable Wastewater Management Practices:

Implement sustainable wastewater management practices to minimize pollution, conserve resources, and protect ecosystems.

Promote water conservation and efficiency measures to reduce wastewater generation and lower water usage in households, industries, and agriculture.

Invest in decentralized wastewater treatment systems such as onsite septic systems, constructed wetlands, or decentralized treatment plants to serve small communities and rural areas where centralized infrastructure may be impractical or cost-prohibitive.

Explore innovative wastewater treatment technologies such as membrane bioreactors, advanced oxidation processes, or anaerobic digestion to improve treatment efficiency, recover valuable resources, and reduce energy consumption and carbon footprint.

Reuse and Recycling of Treated Wastewater:

Encourage the reuse and recycling of treated wastewater for non-potable purposes such as landscape irrigation, industrial processes, and toilet flushing.

Implement water reuse schemes and dual plumbing systems to separate treated wastewater from potable water sources and distribute reclaimed water for appropriate uses.

Educate communities and stakeholders about the safety, benefits, and best practices of water reuse to overcome stigma and misconceptions associated with recycled water.

Regulatory Compliance and Public Engagement:

Ensure compliance with regulatory standards, guidelines, and permits for wastewater treatment and disposal to protect public health and environmental quality.

Foster public awareness, participation, and engagement in wastewater management initiatives through outreach, education, and community involvement.

Collaborate with government agencies, non-profit organizations, industry partners, and community stakeholders to develop and implement integrated wastewater management plans that address local needs and priorities.

By adopting environmentally responsible practices for managing wastewater and sewage, communities can protect water resources, reduce pollution, and promote sustainable development. Remember to prioritize collaboration, innovation, and continuous improvement in wastewater management efforts, and empower individuals and communities to take proactive steps towards a cleaner, healthier, and more sustainable future for all.

# Chapter: Land Use and Sustainable Gardening

Maximizing the Productivity of Your Land for Food Production and Self-Sufficiency

In this chapter, we will explore the principles of land use and sustainable gardening practices to maximize the productivity of your land for food production and achieve greater self-sufficiency. Whether you have a small backyard, a rural homestead, or a community garden plot, strategic land management and sustainable gardening techniques can help you grow healthy, nutritious food, reduce environmental impact, and promote resilience in your local food system.

Assessing Your Land and Resources:

Begin by assessing your land's characteristics, including soil type, sunlight exposure, drainage patterns, and microclimates, to determine the most suitable areas for gardening and food production.
Consider factors such as available water sources, access to tools and equipment, and proximity to urban centers or markets when planning your gardening activities.
Conduct soil tests to evaluate soil fertility, pH levels, and nutrient content, and amend soil as needed to optimize growing conditions for plants.
Planning Your Garden Layout:

Design a garden layout that maximizes space utilization, promotes efficient crop rotation, and minimizes competition for resources such as sunlight, water, and nutrients.
Use companion planting techniques to create beneficial relationships between different plant species, such as attracting pollinators, repelling pests, and enhancing soil fertility.
Incorporate features such as raised beds, trellises, and vertical gardens to optimize growing space and improve accessibility for planting, maintenance, and harvest.
Implementing Sustainable Gardening Practices:

Practice organic gardening methods to minimize the use of synthetic chemicals, pesticides, and fertilizers, and promote soil health, biodiversity, and ecosystem resilience.

Use mulching techniques to suppress weeds, retain soil moisture, and regulate soil temperature, while also adding organic matter and nutrients to the soil as the mulch breaks down.

Practice water-efficient irrigation methods such as drip irrigation, soaker hoses, or rainwater harvesting systems to minimize water waste and promote efficient water use in the garden.

Employ integrated pest management (IPM) strategies to control pests and diseases using natural predators, biological controls, cultural practices, and resistant plant varieties, while minimizing harm to beneficial insects and wildlife.

Growing Food Year-Round:

Extend your growing season and increase food production by incorporating season extension techniques such as cold frames, hoop houses, row covers, or polytunnels to protect plants from frost and extend harvests into cooler months.

Experiment with cool-season crops, succession planting, and intercropping to maintain a steady supply of fresh produce throughout the year, even in regions with short growing seasons or unpredictable weather patterns.

Preserve surplus harvests through techniques such as canning, freezing, drying, or fermenting to enjoy homegrown food year-round and reduce food waste.

Engaging in Community Gardening and Food Sharing:

Participate in community gardening initiatives, cooperative gardens, or urban agriculture projects to share resources, knowledge, and surplus produce with neighbors, friends, and community members.

Collaborate with local organizations, schools, or community groups to establish community gardens, food forests, or community-supported agriculture (CSA) programs to increase access to fresh, healthy food and promote food security and resilience in your community.

Share seeds, seedlings, and gardening tips with fellow gardeners, participate in seed swaps or plant exchanges, and support initiatives that promote seed saving, heirloom varieties, and biodiversity conservation.

By implementing land use strategies and sustainable gardening practices, you can maximize the productivity of your land for food production, reduce environmental impact, and promote self-sufficiency and resilience in your household and community. Remember to prioritize soil health, biodiversity, and resource efficiency in your gardening efforts, and embrace the principles of sustainability and stewardship in your land management practices. With creativity, dedication, and a commitment to sustainable living, you can cultivate a thriving garden that nourishes both body and soul while contributing to a healthier, more resilient food system for all.

# Chapter: Principles of Sustainable Gardening, Permaculture, and Organic Farming

In this chapter, we will explore the principles of sustainable gardening, permaculture, and organic farming, which prioritize environmental stewardship, biodiversity conservation, and resilience in food production systems. By embracing these principles, gardeners and farmers can cultivate healthy, productive ecosystems while minimizing environmental impact and promoting sustainable livelihoods.

Sustainable Gardening Principles:

Embrace biodiversity by planting a diverse range of crops, flowers, and native plants to support pollinators, beneficial insects, and soil microorganisms.
Practice soil conservation techniques such as mulching, cover cropping, and minimal tillage to protect soil structure, fertility, and moisture retention.
Conserve water resources through rainwater harvesting, water-efficient irrigation methods, and drought-tolerant plant selections to minimize water waste and promote efficient water use.
Use organic gardening methods to avoid synthetic chemicals, pesticides, and fertilizers, and promote soil health, biodiversity, and ecosystem resilience.
Prioritize energy efficiency, resource conservation, and waste reduction in garden design, construction, and maintenance practices to minimize environmental impact and carbon footprint.
Permaculture Principles:

Design systems that mimic natural ecosystems, integrate diverse elements, and maximize resource efficiency, productivity, and resilience.
Observe and interact with the natural environment to understand patterns, cycles, and relationships, and design landscapes and food systems that work in harmony with nature.

Use thoughtful design techniques such as zoning, sector analysis, and stacking functions to optimize space utilization, minimize inputs, and enhance ecosystem services.

Emphasize diversity, redundancy, and self-regulation in system design to increase resilience, adaptability, and stability in the face of environmental change and disturbance.

Practice thoughtful stewardship and ethical decision-making to care for the earth, care for people, and promote fair share and redistribution of resources within communities.

Organic Farming Principles:

Focus on soil health as the foundation of organic farming, utilizing practices such as crop rotation, composting, and green manures to build soil fertility, structure, and microbial activity.

Avoid synthetic chemicals, pesticides, and fertilizers, and instead rely on natural inputs, biological controls, and integrated pest management (IPM) strategies to manage pests and diseases.

Protect water quality and ecosystem health by minimizing runoff, erosion, and contamination from agricultural activities, and promoting sustainable water management practices such as irrigation efficiency and watershed protection.

Support biodiversity and habitat conservation by providing habitat for beneficial insects, birds, and wildlife, and preserving natural areas and hedgerows within and around farm landscapes.

Prioritize regenerative practices such as agroforestry, silvopasture, and perennial polycultures to enhance soil carbon sequestration, mitigate climate change, and promote long-term sustainability and resilience.

By embracing the principles of sustainable gardening, permaculture, and organic farming, gardeners and farmers can cultivate productive, resilient, and regenerative food systems that nourish people, communities, and the planet. Remember to prioritize observation, experimentation, and continuous learning in your gardening and farming practices, and collaborate with nature to create thriving ecosystems that sustain life in all its forms. With dedication, creativity, and a commitment to sustainability, we can cultivate a healthier, more resilient food system for generations to come.

# Chapter: Creating a Resilient Garden that Can Withstand Various Environmental Challenges

In this chapter, we will explore strategies for creating a resilient garden that can thrive despite various environmental challenges such as extreme weather events, pests, diseases, and soil degradation. By implementing resilient gardening practices, gardeners can build healthy, productive ecosystems that are better equipped to withstand and recover from adversity, ensuring a consistent and abundant harvest year after year.

Choose Resilient Plant Varieties:

Select plant varieties that are well-adapted to your local climate, soil conditions, and growing conditions, and are known for their resilience to pests, diseases, and environmental stressors.
Choose heirloom and open-pollinated varieties that have a long history of adaptability and genetic diversity, and experiment with regionally adapted cultivars and native plants that are naturally suited to your area.
Build Healthy Soil:

Prioritize soil health as the foundation of a resilient garden, focusing on practices that improve soil structure, fertility, and microbial activity.

Incorporate organic matter such as compost, manure, and cover crops into the soil to increase organic content, improve water retention, and enhance nutrient availability.
Practice minimal tillage and avoid compaction to protect soil structure and promote beneficial soil organisms such as earthworms, fungi, and bacteria.
Implement Water-Wise Techniques:

Develop water-efficient irrigation systems such as drip irrigation, soaker hoses, or rainwater harvesting systems to minimize water waste and promote efficient water use in the garden.
Mulch garden beds with organic materials such as straw, wood chips, or shredded leaves to retain soil moisture, suppress weeds, and regulate soil temperature, reducing the need for frequent watering.
Practice Integrated Pest Management (IPM):

Adopt an integrated approach to pest management that combines cultural, biological, and mechanical controls to minimize pest populations and reduce reliance on chemical pesticides.
Encourage natural predators such as ladybugs, lacewings, and birds to control pest populations, and plant trap crops or sacrificial plants to attract and divert pests away from valuable crops.
Diversify Plantings and Interplanting:

Cultivate a diverse range of plant species in your garden to create habitat for beneficial insects, birds, and wildlife, and reduce the risk of pest and disease outbreaks through natural biodiversity.
Practice interplanting and companion planting techniques to maximize space utilization, enhance soil fertility, and promote mutualistic relationships between different plant species.
Monitor and Respond to Environmental Stressors:

Regularly monitor your garden for signs of environmental stress such as drought, heat stress, or pest infestations, and take proactive measures to address issues before they escalate. Implement adaptive management strategies such as adjusting watering schedules, providing shade or shelter, or using row covers or protective barriers to mitigate environmental stressors and protect vulnerable plants.

Embrace Flexibility and Experimentation:

Be open to experimentation and adaptation in your gardening practices, and be willing to adjust your approach based on changing environmental conditions, observations, and feedback from your garden.

Embrace the concept of resilience as an ongoing process of learning, adaptation, and evolution, and celebrate the resilience of nature in your garden as it responds and adapts to environmental challenges.

By incorporating these strategies for creating a resilient garden, gardeners can cultivate healthy, productive ecosystems that are better equipped to withstand and recover from environmental challenges. Remember to prioritize observation, experimentation, and collaboration with nature in your gardening practices, and cultivate a mindset of resilience and adaptation that will serve you and your garden well in the face of uncertainty and change. With dedication, creativity, and a commitment to sustainability, you can cultivate a garden that thrives in all seasons and conditions, providing nourishment, beauty, and joy for years to come.

# Chapter: Integrating Livestock, Composting, and Natural Pest Control Methods into Your Gardening Practices

In this chapter, we will explore how to integrate livestock, composting, and natural pest control methods into your gardening practices to create a harmonious and sustainable ecosystem. By incorporating these elements, gardeners can improve soil fertility, manage organic waste, and reduce pest populations naturally, enhancing the overall health and productivity of their gardens.

Benefits of Integrating Livestock into Your Garden:

Livestock such as chickens, ducks, or rabbits can provide valuable services to your garden ecosystem, including pest control, weed management, and nutrient cycling.
Chickens and ducks can help control insect pests by foraging for insects, larvae, and weed seeds, while also providing natural fertilizer through their droppings.
Rabbits can be raised in portable pens or tractors to graze on grass and weeds, providing natural weed control and fertilizing garden beds with their manure.
Implementing Composting Systems:

Composting is a natural process of decomposition that transforms organic waste into nutrient-rich compost, which can be used to improve soil fertility, structure, and moisture retention in the garden.

Set up a composting system using a bin, pile, or vermiculture (worm composting) setup to compost kitchen scraps, yard waste, and livestock manure into valuable organic fertilizer.

Layer compostable materials such as kitchen scraps, yard trimmings, straw, leaves, and livestock bedding in alternating layers to create a balanced mix of carbon and nitrogen-rich materials.

Turn or aerate the compost regularly to facilitate decomposition and microbial activity, and monitor moisture levels to ensure proper composting conditions.

Natural Pest Control Methods:

Embrace natural pest control methods to manage pest populations in your garden without the use of synthetic chemicals or pesticides.

Encourage natural predators such as ladybugs, lacewings, birds, and beneficial insects to control pest populations by providing habitat, food sources, and shelter in your garden.

Plant a diverse range of flowering plants, herbs, and companion crops to attract pollinators and beneficial insects, and create habitat for natural predators to thrive.

Use physical barriers, traps, or exclusion techniques to protect vulnerable plants from pests, and practice crop rotation and intercropping to disrupt pest life cycles and reduce pest pressure.

Managing Livestock and Composting:

Integrate livestock management and composting practices to maximize nutrient cycling and soil fertility in your garden.

Use livestock bedding such as straw, hay, or wood shavings as carbon-rich materials in your composting system, along with manure and other organic waste.

Allow chickens or ducks to scratch and turn compost piles, helping to aerate the compost and accelerate decomposition while also controlling insect pests and weed seeds.
Monitoring and Adjusting Practices:

Regularly monitor your garden for signs of pest infestations, nutrient deficiencies, or imbalances, and adjust your practices accordingly to maintain a healthy and productive ecosystem. Be proactive in addressing pest outbreaks or nutrient deficiencies by implementing natural pest control methods, adjusting composting practices, or amending soil as needed.

By integrating livestock, composting, and natural pest control methods into your gardening practices, you can create a balanced and sustainable ecosystem that supports healthy plant growth, reduces reliance on external inputs, and promotes ecological resilience. Remember to prioritize observation, experimentation, and collaboration with nature in your gardening endeavors, and embrace the interconnectedness of all living organisms in your garden ecosystem. With dedication, creativity, and a commitment to sustainability, you can cultivate a thriving garden that nourishes both body and soul while promoting harmony and balance in the natural world.

# Conclusion: Recap of Key Concepts Covered in the Home Defense Book

In the pages of the Home Defense Book, we have explored a comprehensive guide to sustainable living, emergency preparedness, and resilience-building strategies for homeowners and communities alike. Throughout our journey, we have delved into a wide range of topics, from sustainable gardening practices to renewable energy systems, from emergency preparedness plans to natural disaster management. Here, we recap the key concepts covered in this indispensable guide:

Sustainable Living Practices:

Understanding the importance of sustainable living in modern times and the benefits of reducing waste, conserving energy, and minimizing environmental impact.
Exploring eco-friendly habits such as recycling, composting, and using energy-efficient appliances to promote sustainability at home.
Electrical Systems and Power Tools:

Learning about electrical safety precautions, basics of wiring and circuitry, and troubleshooting techniques for homeowners.

Guide to selecting and using power tools safely and effectively for DIY projects and home maintenance.

Power Systems and Energy Independence:

Introduction to renewable energy sources such as solar, wind, and hydro power, and their role in achieving energy independence.

Setting up and maintaining home-based renewable energy systems to reduce reliance on traditional power grids.

Survival and Emergency Preparedness:

Importance of being prepared for emergencies and disasters, and creating comprehensive emergency preparedness plans for homes and families.

Essential survival skills including first aid, fire starting, shelter building, and foraging, as well as tips for stockpiling emergency supplies and creating survival kits.

Handling Natural Disasters:

Understanding different types of natural disasters and their potential impact, and preparing for earthquakes, hurricanes, wildfires, floods, and other calamities.

Evacuation planning, sheltering in place, and community response strategies for managing natural disasters effectively.

Water Supply and Management:

Importance of clean water for survival and sustainability, and strategies for managing water supply, conserving water, and ensuring water quality at home.

Rainwater harvesting, water purification techniques, and water conservation strategies for sustainable water management.

Land Use and Sustainable Gardening:

Maximizing the productivity of land for food production and self-sufficiency through sustainable gardening practices, permaculture, and organic farming.

Creating a resilient garden that can withstand various environmental challenges by integrating livestock, composting, and natural pest control methods.

In closing, the Home Defense Book equips readers with the knowledge, skills, and resources needed to create a safe, sustainable, and resilient home environment. By embracing the principles and practices outlined in this book, readers can not only protect their homes and families but also contribute to a more sustainable and resilient future for themselves and their communities. Whether preparing for emergencies, managing natural resources, or cultivating healthy gardens, the Home Defense Book empowers readers to take proactive steps towards self-reliance, sustainability, and preparedness in an ever-changing world.

# Chapter: Encouragement to Take Action and Implement Sustainable Living Practices, Emergency Preparedness Measures, and Survival Skills

**Dear Readers,**

As you reach the end of the Home Defense Book, I want to offer you a heartfelt encouragement to take action and implement the valuable knowledge and skills you've gained from these pages. The concepts we've explored - from sustainable living practices to emergency preparedness measures and survival skills - are not merely theoretical; they are practical tools that can empower you to protect yourself, your loved ones, and your home in the face of adversity.

Sustainable Living Practices:
It's time to embrace sustainable living practices in your daily life. Start small by reducing waste, conserving energy, and minimizing your environmental impact. Whether it's recycling, composting, or switching to energy-efficient appliances, every action you take makes a difference. Challenge yourself to adopt one new sustainable habit each week and watch as these small changes add up to significant positive impact over time.

Emergency Preparedness Measures:
Don't wait until disaster strikes to prepare yourself and your family. Create a comprehensive emergency preparedness plan tailored to your unique circumstances and needs. Stockpile emergency supplies, establish communication protocols, and designate evacuation routes. Practice emergency drills with your family so that everyone knows what to do in case of an emergency. Remember, preparedness is key to staying safe and resilient in challenging times.

Survival Skills:
Equip yourself with essential survival skills that can mean the difference between life and death in an emergency situation. Learn first aid, fire starting, shelter building, and foraging techniques. Practice these skills regularly until they become second nature. Remember, knowledge is power, and the more prepared you are, the better equipped you'll be to navigate unexpected challenges.

Taking Action:
Now is the time to take action. Don't let fear or procrastination hold you back. Start implementing the sustainable living practices, emergency preparedness measures, and survival skills you've learned from this book today. Share your knowledge with friends, family, and neighbors, and encourage them to join you on this journey towards self-reliance and resilience.

In Conclusion:
I believe in your ability to create a safe, sustainable, and resilient home environment. By taking proactive steps to implement the principles and practices outlined in this book, you are not only protecting yourself and your loved ones but also contributing to a more sustainable and resilient future for all. Together, we can build a world where every home is a sanctuary of safety, preparedness, and sustainability.

With courage, determination, and a commitment to action, you can make a positive difference in your life and in the world around you. Remember, the power to create change lies within you. Seize the opportunity to take action and make a lasting impact today.

**Wishing you strength, resilience, and success on your journey,**
**Laurel D. Malvern**

# Chapter: Inspiring Stories of Home Transformation into Resilient, Self-Sufficient Havens

In this chapter, we will delve into the inspiring stories of individuals who have transformed their homes into resilient, self-sufficient havens. These stories demonstrate the power of determination, creativity, and community collaboration in creating sustainable, prepared, and thriving living spaces that serve as models of resilience and inspiration for others.

The Urban Homesteader:
Sarah, a resident of a bustling urban neighborhood, turned her small backyard into a thriving urban homestead. With limited space but boundless determination, she converted unused lawn space into raised beds for vegetable gardening, installed a rainwater harvesting system to collect water for irrigation, and built a small coop for raising chickens. Through her dedication to sustainable living practices and permaculture principles, Sarah not only produces a significant portion of her family's food but also inspires her neighbors to adopt similar practices, creating a greener and more resilient community.

The Off-Grid Pioneer:

John and Emily, a couple living in a remote rural area, decided to embrace off-grid living as a way to reduce their environmental footprint and increase their self-sufficiency. They installed solar panels for electricity, dug a well for water supply, and built a greenhouse for year-round food production. Despite the challenges of living off-grid, including harsh weather conditions and limited access to resources, John and Emily have thrived in their self-sufficient lifestyle, proving that with determination and resourcefulness, off-grid living is not only possible but also deeply rewarding.

The Community Garden Champion:
Mark, a passionate advocate for community gardening, spearheaded the creation of a community garden in his neighborhood. Through grassroots organizing and collaboration with local residents and organizations, he transformed an abandoned lot into a vibrant community hub for gardening, education, and social interaction. The community garden not only provides fresh produce for participants but also fosters a sense of belonging, connection, and empowerment among neighbors. Mark's vision and leadership have transformed the neighborhood, demonstrating the power of community-driven initiatives in creating resilient and thriving communities.

The Sustainable Suburban Family:

The Johnson family, residents of a suburban neighborhood, embarked on a journey towards sustainable living and resilience after experiencing the impacts of a severe storm that left their community without power for several days. Determined to be better prepared for future emergencies, they installed a backup generator, built raised beds for vegetable gardening, and installed a rainwater collection system. Through their commitment to preparedness and sustainability, the Johnson family has not only enhanced their resilience to emergencies but also fostered a deeper connection to their environment and community.

These stories of home transformation serve as powerful reminders of the potential for individuals and communities to create resilient, self-sufficient havens that promote sustainability, preparedness, and community resilience. By sharing their experiences and inspiring others to take action, these individuals demonstrate that even small-scale changes can have a significant impact on building a more sustainable and resilient future for all.

# Chapter: Closing Thoughts from the Author

**Dear Readers,**

As we come to the end of the Home Defense Book, I want to leave you with some closing thoughts that emphasize the importance of being proactive in safeguarding ourselves and the planet. Throughout this journey, we've explored a multitude of strategies and practices aimed at creating resilient, self-sufficient homes and communities. But beyond the practical tips and techniques lies a deeper message - a call to action to take responsibility for our own well-being and the health of our planet.

In today's rapidly changing world, with climate change, environmental degradation, and societal challenges on the rise, it's more crucial than ever to be proactive in safeguarding ourselves and the planet. We cannot afford to wait for others to take action or for crises to unfold before we act. Instead, we must embrace a proactive mindset that empowers us to anticipate challenges, mitigate risks, and build resilience in our homes, communities, and beyond.

Being initiative-taking means taking initiative to prepare for emergencies, whether it is creating an emergency preparedness plan, stockpiling supplies, or learning essential survival skills. It means adopting sustainable living practices that reduce our environmental footprint and promote resilience in the face of environmental challenges. It means fostering strong, connected communities that support one another in times of need and work together to address shared challenges.

At its core, being proactive is about recognizing our interconnectedness - with each other and with the natural world - and taking responsibility for our collective future. It's about acknowledging that the choices we make today have far-reaching consequences for future generations and the health of the planet. By being proactive in safeguarding ourselves and the planet, we not only protect our own well-being but also contribute to a more sustainable, resilient, and equitable world for all.

As you close the pages of this book and embark on your journey towards greater self-sufficiency, preparedness, and sustainability, I urge you to carry forward the spirit of proactivity and resilience in everything you do. Be initiative-taking in seeking out knowledge, in acting, and in building connections with others who share your vision for a better world. Together, we have the power to create positive change and build a brighter future for ourselves and future generations.

Thank you for joining me on this journey. May your homes be havens of safety, sustainability, and resilience, and may your actions inspire others to follow suit.

**With gratitude and hope,**

**Laurel D. Malvern**
**Author, Home Defense Book**

www.ingramcontent.com/pod-product-compliance
Lightning Source LLC
Chambersburg PA
CBHW050325230526
45471CB00005B/2355